Thermometers

by Stephen Schutz

Starfall Education
P.O. BOX 359, Boulder, CO 80306

Copyright © 2007, 2012 by Starfall Education. All rights reserved. Starfall® is a registered trademark in the U.S., the European Community and various other countries. Printed in China.
ISBN: 978-1-59577-062-2

How **Hot**?

How Cold?

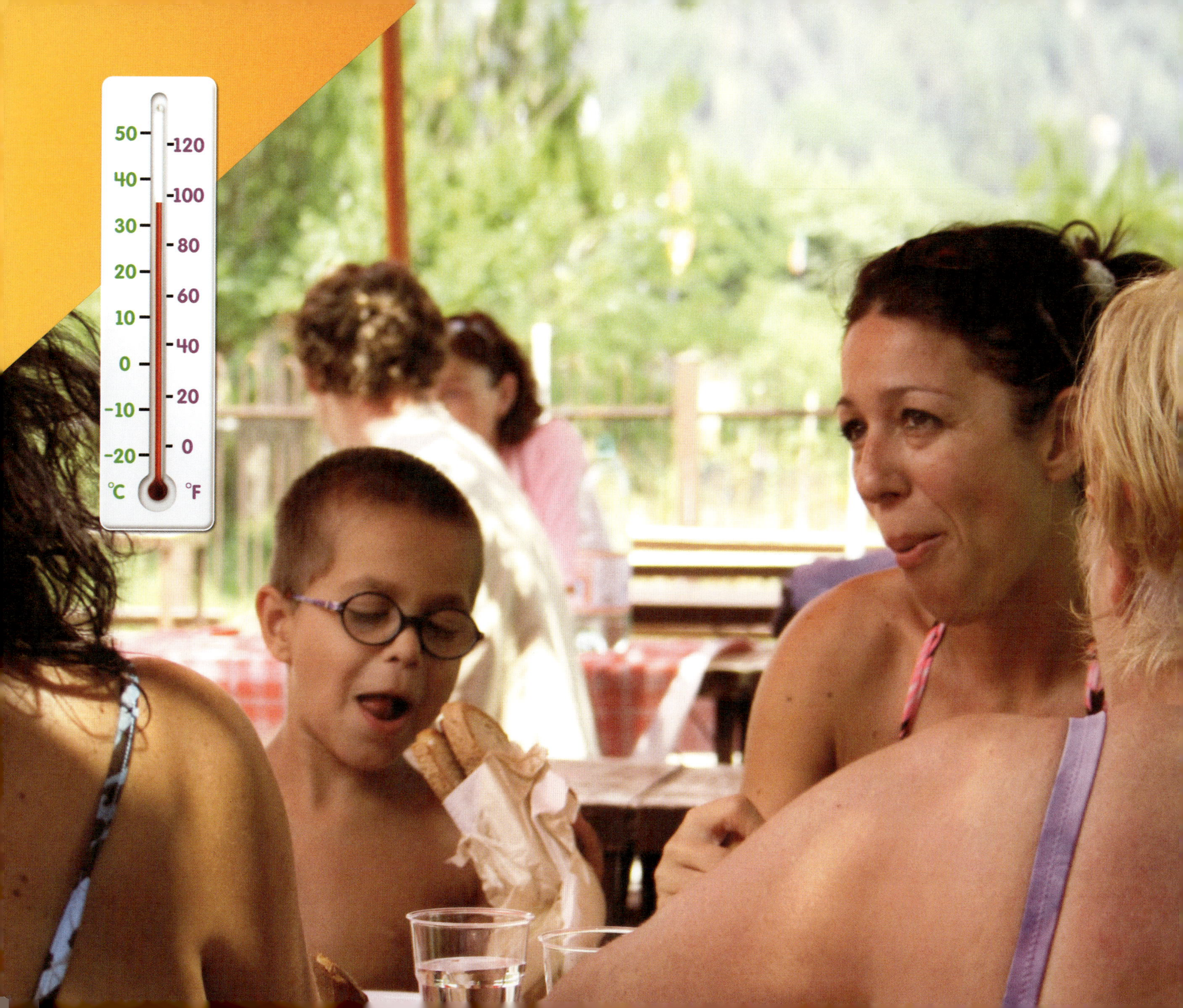
50
40
30
20
10
0
-10
-20
°C
120
100
80
60
40
20
0
°F

You can feel hot.

A **thermometer** tells you *how* hot.

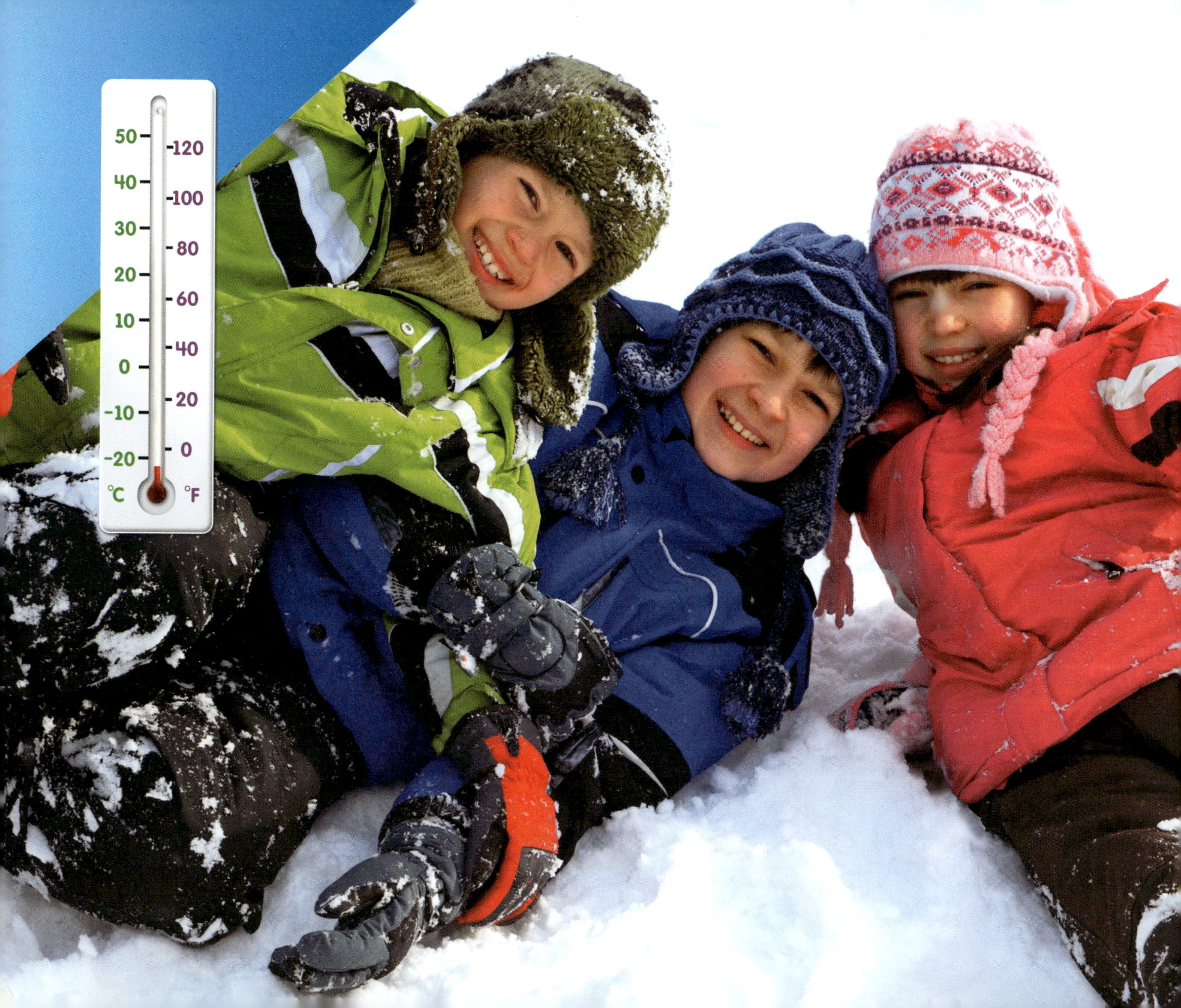
50
40
30
20
10
0
-10
-20
°C
120
100
80
60
40
20
0
°F

You can feel cold.

A thermometer tells you *how* cold.

Hot

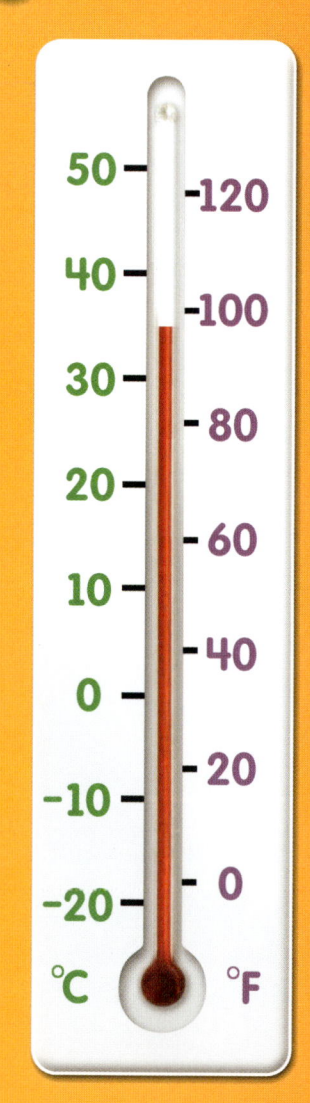

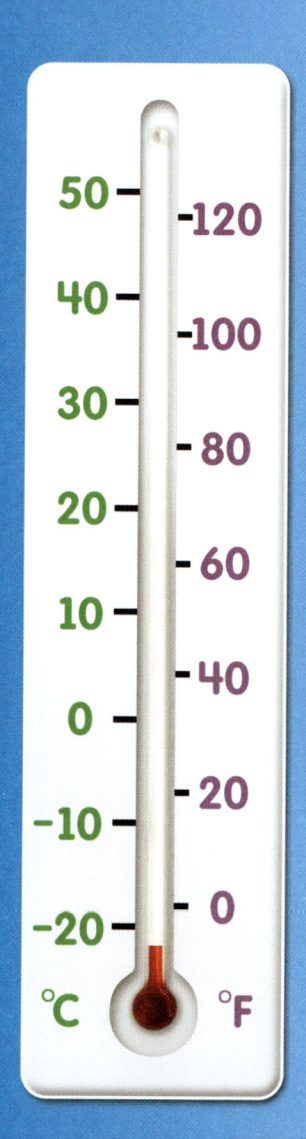

Cold

Thermometers tell us how hot or how cold.

This is the **temperature**.

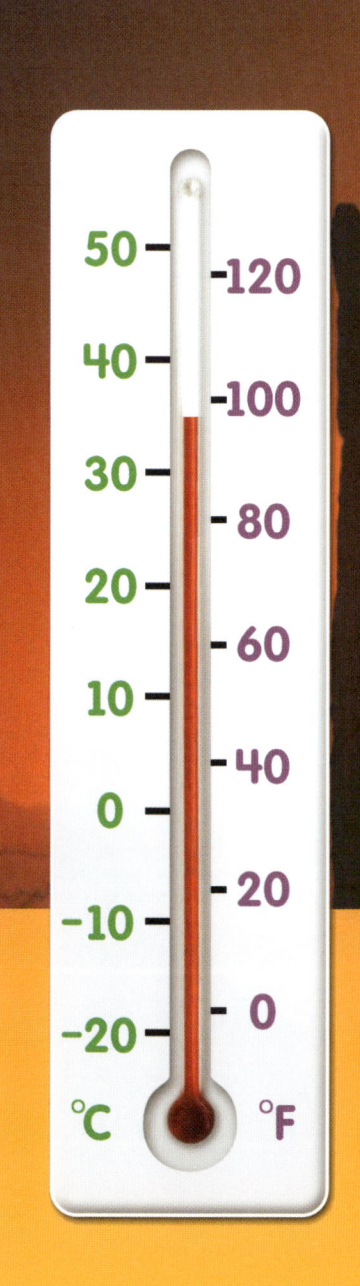

High

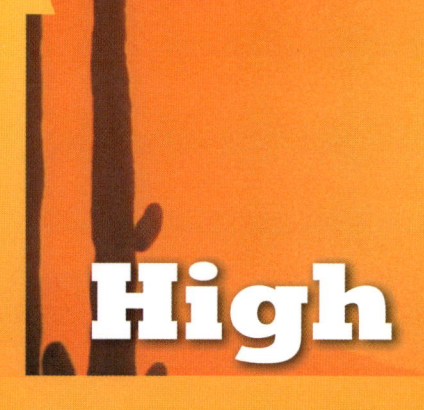

When it is hot, the line goes up.

The temperature is high.

Low

When it is cold, the line goes down.

The temperature is low.

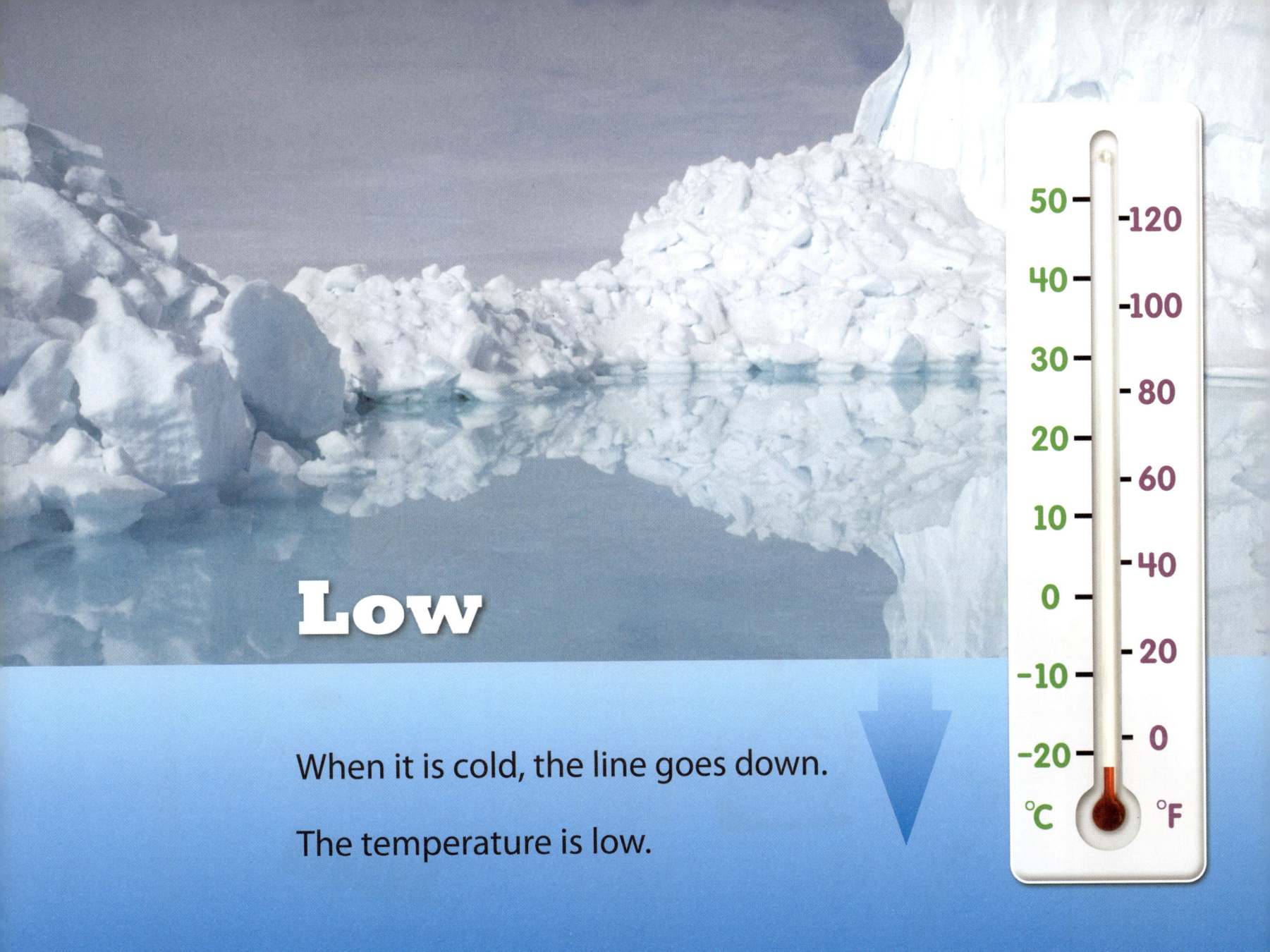

°C
°F
-60
-40
-20
0
20
40
60
80
100
120
-50
-40
-30
-20
-10
0
10
20
30
40
50
°C
°F
50
40
30
20
10
0
-10
-20
120
100
80
60
40
20
0

Roast
Oven
Temperature
100
150
200
250
300
350
400
450
500
550
600
60
80
100
120
140
160
180
200
220
°F

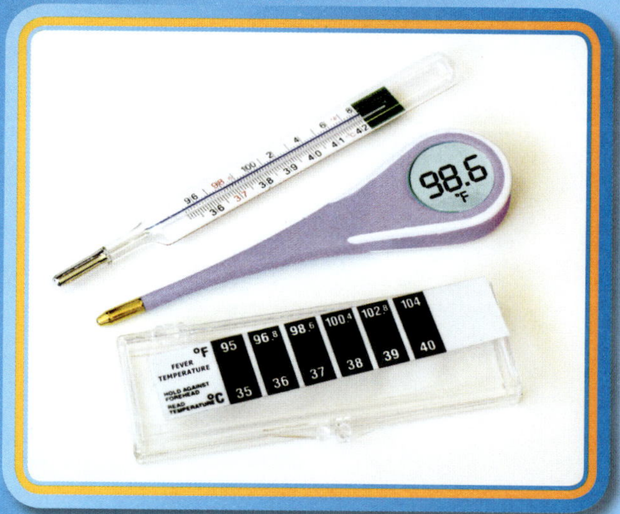
98.6
°F
FEVER TEMPERATURE
HOLD AGAINST FOREHEAD
READ TEMPERATURE
°F
°C
95
96.8
98.6
100.4
102.8
104
35
36
37
38
39
40

There are many kinds of thermometers.

All thermometers tell us the temperature.
They tell us how hot and how cold.

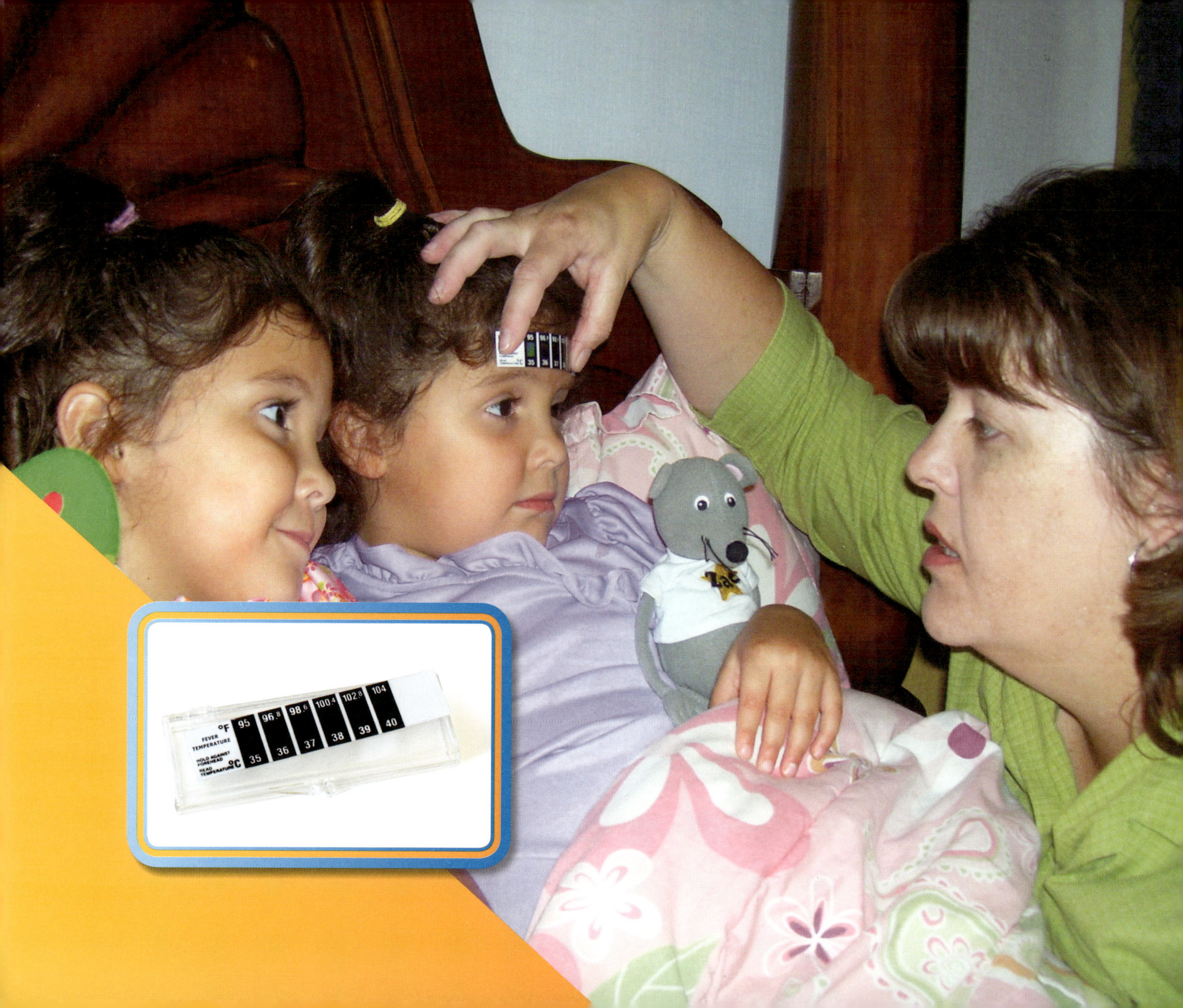
°F
95
96.8
98.6
100.4
102.8
104
FEVER TEMPERATURE
HOLD AGAINST FOREHEAD
HEAD TEMPERATURE
°C
35
36
37
38
39
40

I feel sick! Am I hot?

Mom uses this thermometer to see.

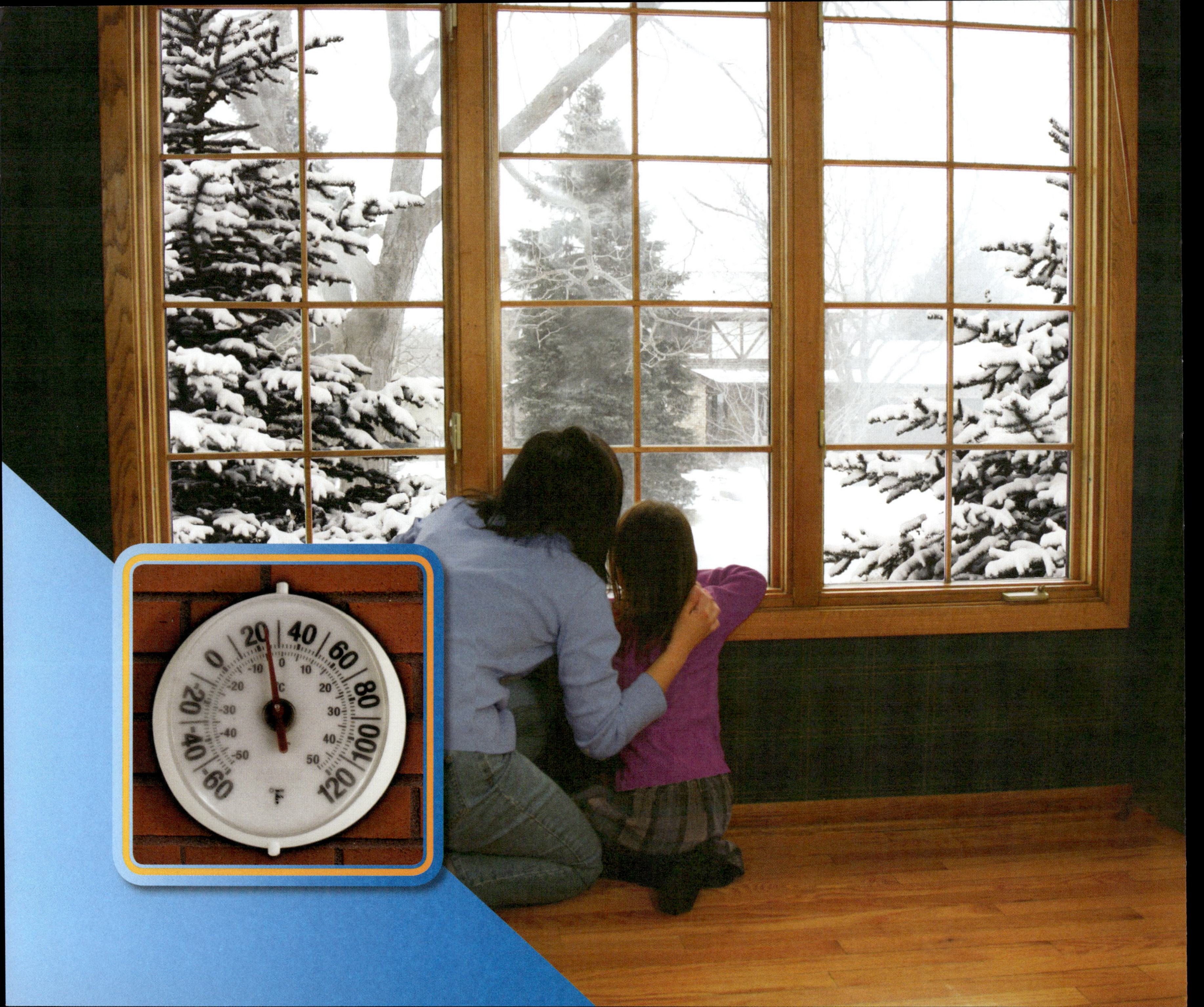

-60 -40 -20 0 20 40 60 80 100 120
°F
-50 -40 -30 -20 -10 0 10 20 30 40 50
°C

It looks cold today! Will I need a coat?

I use this thermometer to see.

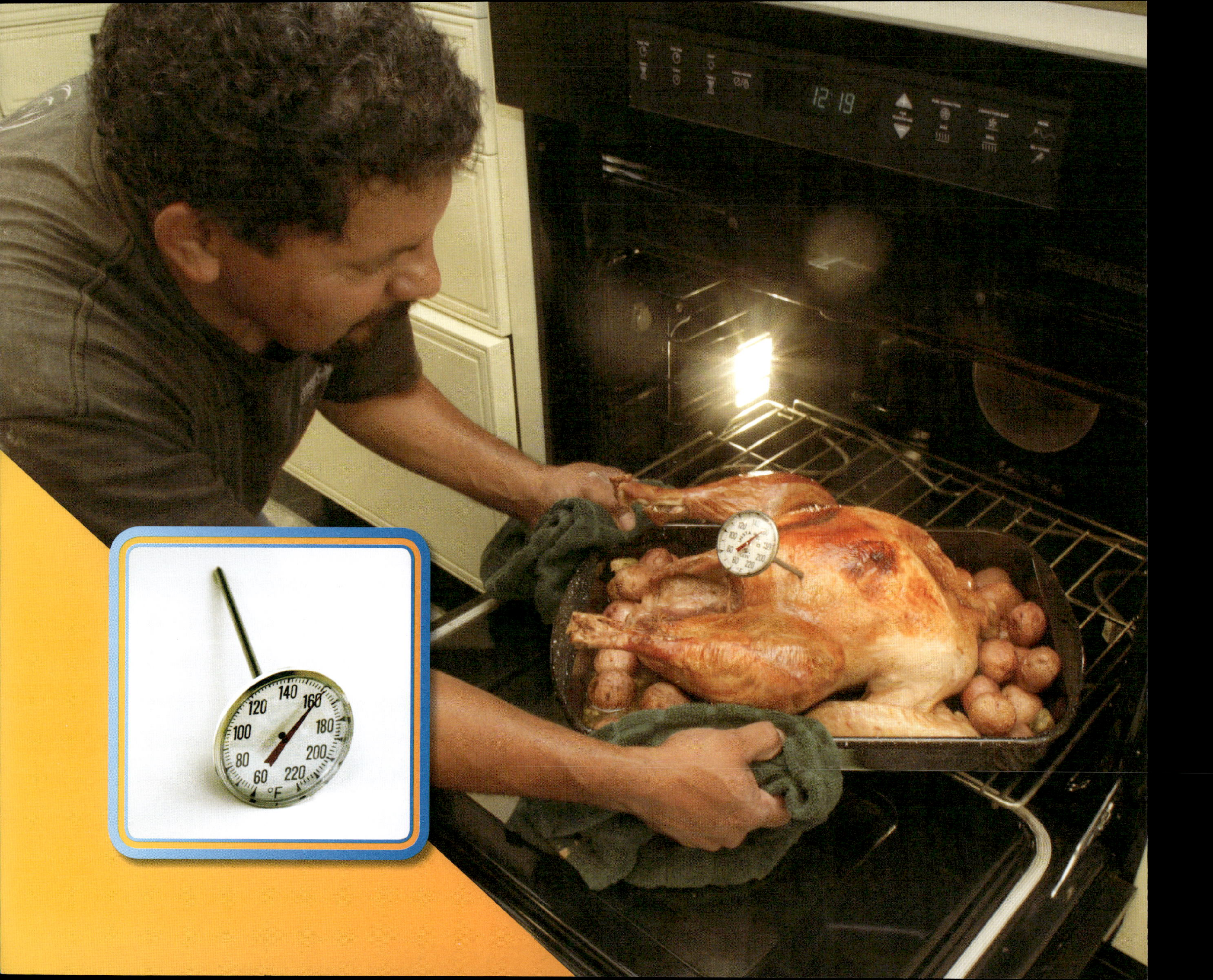
12 19
120
140
160
100
180
80
200
60
220
°F

That looks yummy! Can we eat it?

Dad uses this thermometer to see.

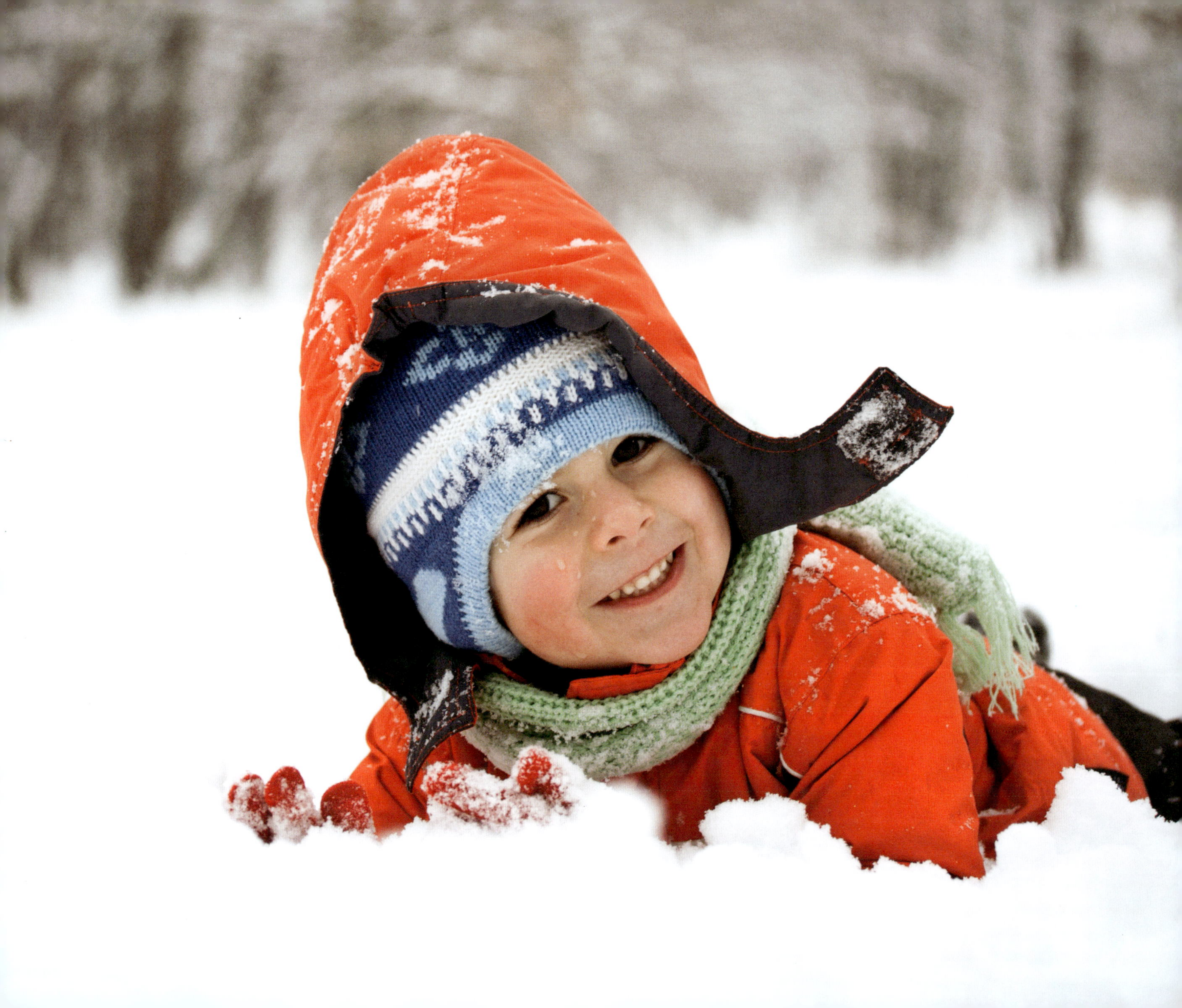

Look at this picture. Is the temperature high or low?

What do you think the thermometer will say?

Words You Know

Temperature The measure of how hot or how cold.
Thermometer An instrument that measures temperature.

Thermometers

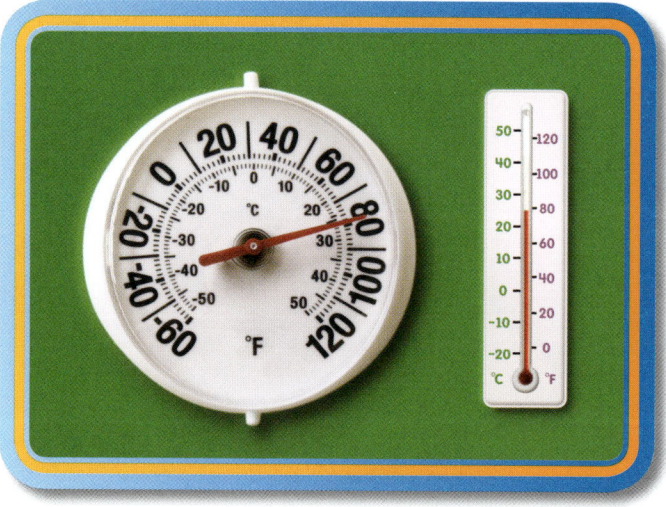

We use different thermometers for different purposes. Look at the groups of thermometers above. Do you know what each one is used for? Which group of thermometers will tell you if you have a fever? Which would you use for cooking? Which would you use to check the air?

Learn More About Thermometers

Celsius and Fahrenheit

We measure temperature in degrees. There are two sets of numbers (or scales) used to describe degrees, Fahrenheit and Celsius.

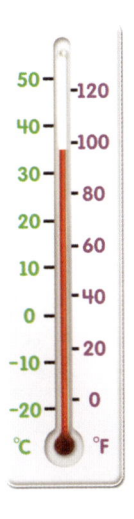

Look at this thermometer. The purple numbers on the right have an F° below them. The little ° next to the F means degrees. This side of the thermometer measures the temperature in Fahrenheit degrees.

The green numbers on the left have a C° below them. This side of the thermometer measures the temperature in Celsius degrees.

The Celsius scale is the standard scale used around most of the world to measure temperatures. Only the United States continues to use the Fahrenheit scale.

How does a thermometer work?

All thermometers have something inside of them that responds to temperature changes in some way.

This thermometer has a red liquid inside of a glass tube. As the temperature rises, the liquid expands and the red line goes up!

This thermometer has a coil-spring inside of it. As the temperature rises, the spring expands and pushes the dial to the right.

This thermometer has a small electrical current inside that changes with temperature.

Index

C
Cold **3**, **7**, **9**, **11**, **13**, **17**, **22**

H
High **10**, **21**

Hot **2**, **5**, **9**, **10**, **13**, **15**, **22**

L
Low **11**, **21**

T
Temperature **9**, **10**, **11**, **13**, **21**, **22**, **23**

Thermometer **5**, **7**, **9**, **13**, **15**, **17**, **19**, **21**, **22**, **23**

About the Author

At age 9, young Stephen Schutz was still struggling to read. What came easily for some children required many more hours of Stephen's work, and he was consistently toward the bottom of his class in reading. Now a Ph.D. in physics and a successful publisher and artist, Dr. Schutz wants to make sure children in his situation today have resources that can help. He turned to the Internet and conceived Starfall.com, an online educational resource available to children the world over.

Acknowledgements

Developed in cooperation with ELHI Publishers Services. The following photos were licensed from iStockphoto: *Siblings in the snow*, on page 6, © Marzanna Syncerz; *Little happy boy*, on page 20, © grafvision.